I Get It!

An MCP MATH Series

ADDITION AND SUBTRACTION

Modern Curriculum Press

Parsippany, New Jersey

WRITER
Barbara Bando Irvin

EDITOR
Harriet Slonim

EXECUTIVE EDITOR
Jill Levy

PRODUCTION COORDINATOR
Alan Dalgleish

CONTENT REVIEWERS

Mattie Adelaja
Title I Coordinator
Golfcrest Elementary School
Houston, Texas

David J. Glatzer
Supervisor of Mathematics
West Orange Public Schools
West Orange, New Jersey

Joyce Glatzer
Mathematics Consultant
West Paterson, New Jersey

Bobbi Defenbaugh
Mathematics Consultant
Alexandria, Virginia

Dr. Gwendolyn E. Long
Principal, John Farren Fine Arts Elementary School
Chicago, Illinois

DESIGNERS
Evelyn O'Shea, Greg Dulles, Bernadette Hruby, Denise Ingrassia,
Dorothea Fox, Liz Kril

ART BUYER
Daniel Trush

ILLUSTRATIONS
P. T. Pie

PHOTOS
All Photographs © Pearson Learning unless otherwise noted.
63: New York Convention & Visitors Bureau.

Modern Curriculum Press
An imprint of Pearson Learning
299 Jefferson Road, P.O. Box 480
Parsippany, NJ 07054-0480

www.pearsonlearning.com

1-800-321-3106

2 3 4 5 6 7 8 9 10 PO 10 09 08 07 06 05 04 03 02 01

NOTE TO KIDS

This book is all about addition and subtraction. You probably add and subtract more often than you think. When was the last time you had to find out how many of something you had? When did you last have to decide if one thing cost more than another? Each of these times, you probably had to add or subtract—maybe even both.

One of the best things about this book is that it will teach you ways to remember basic facts quickly. You will also learn to add and subtract "in your head." You'll see why you don't always need a pencil and paper or a calculator. Quick—what's $46 + 59$? $534 + 420$? $5,000 - 8$? Wouldn't it be great to be able to solve these in just a few seconds? And, for those times when you don't need an exact answer, you will learn how to make a good guess, an *estimate* of the answer. Estimates come in handy in school, at home, and wherever else you go!

As you do the work in this book, it's a good idea to keep a math journal. Jot down your math ideas in it. Share what you write about addition and subtraction with your class, your teacher, and your family. By the time you finish this book, you will be adding and subtracting numbers in the thousands!

ADDITION AND SUBTRACTION

ADDITION FACTS AND STRATEGIES

SUBTRACTION FACTS AND STRATEGIES

ADDING GREATER NUMBERS

SUBTRACTING GREATER NUMBERS

FIGURE FACTS FAST!

Here are two **strategies,** or ways, to help you learn and remember addition facts.

Count on to add 1, 2, or 3 to a number. Start with the greater number.

3 + 6 = ?

> Think 6.
> Count on 7, 8, 9.

A number line can help.

$$3 + 6 = 9$$

addend   addend   sum

The numbers you add are the **addends.**

The answer is the **sum.**

Use doubles to add numbers that are close to each other. Two addends that are the same form a double.

6 + 8 = ?

> Think 6 + 6 = 12.
> Then add 2 more.

Since 8 is 2 more than 6, then 6 + 8 is 2 more than 12.

So 6 + 8 = 14.

Your Turn

Count on to find each sum.

1. 9 + 2 = _____

2. 3 + 8 = _____

3. 7 + 2 = _____

4. 3 + 7 = _____

5. 6 + 2 = _____

6. 1 + 8 = _____

7. 6 + 2 = _____

8. 2 + 7 = _____

9. 3 + 8 = _____

10.   2
        + 5

11.   6
        + 3

12.   1
        + 7

13.   3
        + 4

14.   7
        + 3

15.   2
        + 8

Use a double to find each sum. Write the double fact you used.

16. 5 + 6 = ____

____ + ____ = ____

17. 7 + 6 = ____

____ + ____ = ____

18. 7 + 9 = ____

____ + ____ = ____

19. 8 + 6 = ____

____ + ____ = ____

20. 5 + 7 = ____

____ + ____ = ____

21. 8 + 9 = ____

____ + ____ = ____

Add. Use doubles or count on.

22. 7 + 8 = ____

23. 2 + 9 = ____

24. 6 + 9 = ____

25. 8 + 5 = ____

26. 9 + 7 = ____

27. 2 + 7 = ____

28. Math Journal Show how to count on to find 5 + 3. Then show how to use a double to find the same sum.

★ PROBLEM SOLVING

Use the clock to count on hours. Solve each problem.

29. A concert starts at 1:30 P.M. It is 2 hours long. At what time will it end?

30. A train trip lasts 3 hours. It starts at 11:00 A.M. At what time will it be over?

_____

_____

💡 Critical Thinking

This is how Marty counted on to find 9 + 3. What did Marty do wrong? Explain.

> Start at 9.
> Count 9, 10, 11.
> 9 + 3 = 11

_____

_____

_____

FRAME IT!

Making **10** is another way to learn and remember addition facts. This is a big help when one addend is 7, 8, or 9.

$4 + 8 = ?$

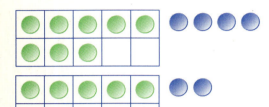

Show the greater number in a ten frame. Show the lesser number outside the frame.

Use part of the lesser number to make 10. Then add the rest.

$8 + 4 = 10 + 2$, so $4 + 8 = 12$.

Your Turn

Draw lines to match the fact with the picture. Write the sum.

1. $8 + 6 =$ _____

2. $7 + 5 =$ _____

3. $5 + 9 =$ _____

Add. Use counters and a ten frame if you like.

4. $9 + 5 =$ _____   5. $7 + 4 =$ _____   6. $5 + 8 =$ _____

7. $9 + 8 =$ _____   8. $3 + 9 =$ _____   9. $8 + 7 =$ _____

10. $6 + 8 =$ _____   11. $9 + 6 =$ _____   12. $8 + 8 =$ _____

REMEMBER:
Use counting on
and doubles too!

Find each sum.

13. 9
    + 2

14. 7
    + 8

15. 4
    + 9

16. 6
    + 5

17. 8
    + 3

18. 6
    + 7

19. 8
    + 4

20. 7
    + 7

21. 4
    + 7

22. 9
    + 4

23. 2
    + 8

24. 4
    + 6

25. 9
    + 9

26. 8
    + 5

27. 5
    + 9

28. 7
    + 9

29. 7
    + 6

Compare. Write <, =, or >. Remember that < means "is less than" and > means "is greater than."

30. $3 + 9 \bigcirc 8 + 2$

31. $1 + 9 \bigcirc 7 + 5$

32. $8 + 8 \bigcirc 7 + 7$

★ PROBLEM SOLVING

Follow the rule. Find the missing numbers.

33.
Add 3.	
4	7
6	
7	
9	

34.
Add the double.	
5	10
7	
4	
9	

35.
Add 9.	
3	12
5	
7	
9	

That's **AMAZING!**

Did you know that you can make any even number odd just by adding 1? Try it. What happens if you add 2 or 3 to any odd number? What pattern do you see?

_____

_____

_____

ORDER, PLEASE!

Sometimes rules can also help you learn and remember addition facts.

Rule 1 You can add numbers in any order. The sum will be the same.

$7 + 4 = 11$ $\qquad\qquad$ $4 + 7 = 11$

Rule 2 When you add zero to a number, the sum is that number.

$5 + 0 = 5$ $\qquad\qquad$ $0 + 5 = 5$

✎ Your Turn

Add.

1. $9 + 0 =$ _____

$0 + 9 =$ _____

2. $6 + 3 =$ _____

$3 + 6 =$ _____

3. $0 + 6 =$ _____

$6 + 0 =$ _____

4. $8 + 5 =$ _____

$5 + 8 =$ _____

5. $0 + 8 =$ _____

$8 + 0 =$ _____

6. $8 + 7 =$ _____

$7 + 8 =$ _____

7.
$$\begin{array}{r} 3 \\ + 9 \\ \hline \end{array} \quad \begin{array}{r} 9 \\ + 3 \\ \hline \end{array}$$

8.
$$\begin{array}{r} 6 \\ + 8 \\ \hline \end{array} \quad \begin{array}{r} 8 \\ + 6 \\ \hline \end{array}$$

9.
$$\begin{array}{r} 0 \\ + 7 \\ \hline \end{array} \quad \begin{array}{r} 7 \\ + 0 \\ \hline \end{array}$$

10.
$$\begin{array}{r} 4 \\ + 6 \\ \hline \end{array} \quad \begin{array}{r} 6 \\ + 4 \\ \hline \end{array}$$

11.
$$\begin{array}{r} 0 \\ + 3 \\ \hline \end{array} \quad \begin{array}{r} 3 \\ + 0 \\ \hline \end{array}$$

12.
$$\begin{array}{r} 9 \\ + 5 \\ \hline \end{array} \quad \begin{array}{r} 5 \\ + 9 \\ \hline \end{array}$$

13.
$$\begin{array}{r} 3 \\ + 5 \\ \hline \end{array} \quad \begin{array}{r} 5 \\ + 3 \\ \hline \end{array}$$

14.
$$\begin{array}{r} 6 \\ + 5 \\ \hline \end{array} \quad \begin{array}{r} 5 \\ + 6 \\ \hline \end{array}$$

15.
$$\begin{array}{r} 4 \\ + 0 \\ \hline \end{array} \quad \begin{array}{r} 0 \\ + 4 \\ \hline \end{array}$$

**Find the sum. Then change the order of the addends and add again.
The first one is done for you.**

16. $4 + 5 =$ ___9___

___$5 + 4 = 9$___

17. $9 + 6 =$ _____

_____

18. $8 + 9 =$ _____

_____

19. $0 + 7 =$ _____

_____

20. $4 + 8 =$ _____

_____

21. $4 + 0 =$ _____

_____

Shade the oval for the correct answer.

22. $9 + 7 =$
(A) 12
(B) 16
(C) 17
(D) 18

23. $6 + 6 =$
(A) 12
(B) 13
(C) 14
(D) 16

24. $8 + 0 =$
(A) 0
(B) 8
(C) 16
(D) 80

Problem Solving

Use what you learned to find each missing number.

25. $8 + \boxed{} = 8$

26. $5 + 7 = 7 + \boxed{}$

27. $0 + 9 = \boxed{} + 0$

28. $1 + \boxed{} = 5 + 1$

29. $\boxed{} + 6 = 6 + 9$

30. $8 + 7 = \boxed{} + 8$

Critical Thinking

**Look for a pattern in each table. Write the rule.
The first one is done for you.**

1. Add 4.

3	7
4	8
5	9
6	10

2.

8	8
7	7
6	6
5	5

3.

4	8
5	10
6	12
7	14

TWO AT A TIME

What if you saw 7 swallows, 2 robins, and 3 blue jays?
How many birds did you see? How can you find out?

You can add of course!

$$
\begin{array}{r} 7 \\ 2 \\ + \ 3 \\ \hline 12 \end{array} \quad 9
$$

$$
\begin{array}{r} 7 \\ 2 \\ + \ 3 \\ \hline \end{array} \quad 5 \quad
\begin{array}{r} + \ 7 \\ \hline 12 \end{array}
$$

$$
\begin{array}{r} 7 \\ 2 \\ + \ 3 \\ \hline \end{array} \quad 10 \quad
\begin{array}{r} + \ 2 \\ \hline 12 \end{array}
$$

TIP
Group numbers that are easy to add. Think of doubles, count on, or make 10.

You saw 12 birds.

✏️ Your Turn

Add.

1. $7 + 2 + 8 =$ _____   **2.** $5 + 5 + 6 =$ _____   **3.** $9 + 1 + 5 =$ _____

4. $3 + 8 + 7 =$ _____   **5.** $8 + 7 + 2 =$ _____   **6.** $6 + 6 + 4 =$ _____

7. $3 + 2 + 0 + 8 =$ _____   **8.** $5 + 0 + 5 + 3 =$ _____   **9.** $5 + 0 + 9 + 1 =$ _____

10. $4 + 4 + 0 + 6 =$ _____   **11.** $7 + 1 + 9 + 3 =$ _____   **12.** $6 + 5 + 0 + 5 =$ _____

13.
$$\begin{array}{r} 4 \\ 3 \\ + \ 3 \\ \hline \end{array}$$
14.
$$\begin{array}{r} 5 \\ 0 \\ + \ 5 \\ \hline \end{array}$$
15.
$$\begin{array}{r} 2 \\ 8 \\ + \ 3 \\ \hline \end{array}$$
16.
$$\begin{array}{r} 3 \\ 7 \\ + \ 4 \\ \hline \end{array}$$
17.
$$\begin{array}{r} 9 \\ 1 \\ + \ 1 \\ \hline \end{array}$$
18.
$$\begin{array}{r} 0 \\ 6 \\ + \ 4 \\ \hline \end{array}$$

19.
$$\begin{array}{r} 3 \\ 2 \\ + \ 8 \\ \hline \end{array}$$
20.
$$\begin{array}{r} 3 \\ 4 \\ 4 \\ + \ 6 \\ \hline \end{array}$$
21.
$$\begin{array}{r} 5 \\ 0 \\ 7 \\ + \ 3 \\ \hline \end{array}$$
22.
$$\begin{array}{r} 1 \\ 2 \\ 3 \\ + \ 5 \\ \hline \end{array}$$
23.
$$\begin{array}{r} 3 \\ 7 \\ 1 \\ + \ 4 \\ \hline \end{array}$$
24.
$$\begin{array}{r} 4 \\ 6 \\ 5 \\ + \ 5 \\ \hline \end{array}$$

Rewrite the numbers in the order that makes them easiest to add. Then find the sum. The first one is started for you.

25. 3 + 5 + 7

$\underline{\quad 3 + 7 + 5 = \quad}$

26. 6 + 1 + 4 + 8

27. 8 + 7 + 3 + 2

28. 7 + 2 + 8

29. 6 + 3 + 6 + 1

30. 8 + 3 + 8 + 3

31. 1 + 8 + 8

32. 5 + 3 + 0 + 5

33. 3 + 9 + 1 + 4

34. Math Journal   Explain how to add 6 + 3 + 7 in your head.

PROBLEM SOLVING

35. Complete the magic square. Each row, column, and diagonal of a magic square must have the same sum. Use the numbers 1, 3, 7, and 9.

36. What is the sum of each row, column, and diagonal in the

magic square? _____

Magic Square

8		6
	5	
4		2

One for Fun!

BULLS-EYE!

Suppose you could throw 3 darts at this dart board.

1. What will your score be if each dart lands in a

different section? _____

2. What is the greatest score possible? _____

3. What is the greatest score possible if only one dart

lands in the center? _____

TAKE 5

PROBLEM SOLVING

1. Fill a **SHAPE**

What numbers belong on each shape? (Hint: The same number is always used on the same shape.)

$\bigcirc + \bigcirc = 16$

$\square + \bigcirc = 12$

2. Oh, It's Nothing!

$\triangle + \triangle = 0$

What number does $\triangle$ stand for? Explain how you know.

3. Balloons for All

Together, Jake and Vanya have 10 balloons. Together, Vanya and Ben have 6. Jake alone has 6. How many balloons does Ben have?

4. Try-Me **Triangle**

The sum of each side of this triangle is 12. The numbers 1, 3, 4, and 6 are missing. Write them where they belong.

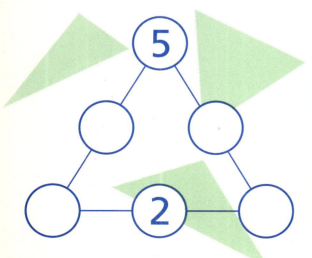

5. Why Make 10?

Salita says that $9 + 7 = 10 + 6$. Do you agree? Explain why or why not.

Make Twenty!

Number of Players: 2 to 4

Goal: To be the first player to get a sum of 20

Game Rules

1 Spin the spinner. Record your number.

2 Spin again. This time add the number you spin to your first number. Write the sum.

3 Players take turns spinning. They add each number they spin to their last numbers to get new sums.

4 The first player to get a sum of exactly 20 wins. If nobody wins, play again!

Materials:
- one spinner divided into four sections labeled 1, 2, 3, 4
- paper and pencil for each player

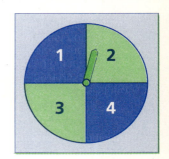

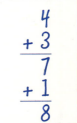

$$\begin{array}{r} 4 \\ +\ 3 \\ \hline 7 \\ +\ 1 \\ \hline 8 \end{array}$$

BACK TRACK!

Count back to subtract 1, 2, or 3 from a number.

9 − 3 = ?

Think 9.
Count back
8, 7, 6.

1 2 3 4 5 6 7 8 ⑨ 10

9 − 3 = 6
↑
difference

Count up to subtract numbers that are close.

10 − 7 = ?

Start at 7.
Count up
8, 9, 10.

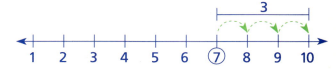

1 2 3 4 5 6 ⑦ 8 9 10

You counted three numbers so 3 is the difference.

10 − 7 = 3
↑
difference

When you subtract, the answer you get is the **difference**.

Your Turn

Count back to find each difference.

1. 9 − 2 = _____

2. 8 − 3 = _____

3. 7 − 2 = _____

4. 12 − 3 = _____

5. 6 − 2 = _____

6. 9 − 1 = _____

7. 7 − 1 = _____

8. 10 − 2 = _____

9. 8 − 2 = _____

10. 10 − 1 = _____

11. 8 − 1 = _____

12. 11 − 2 = _____

Count up to find each difference.

13. 6 − 6 = _____

14. 8 − 6 = _____

15. 7 − 4 = _____

16. 9 − 7 = _____

17. 11 − 9 = _____

18. 12 − 9 = _____

19. 7 − 6 = _____

20. 5 − 4 = _____

21. 6 − 5 = _____

22. 8 − 5 = _____

23. 11 − 8 = _____

24. 10 − 8 = _____

Should you count back or count up? Write the strategy. Then subtract. The first one is started for you.

25.
$$\begin{array}{r} 10 \\ -\ 8 \end{array}$$ count up

26.
$$\begin{array}{r} 5 \\ -\ 2 \end{array}$$ _____

27.
$$\begin{array}{r} 8 \\ -\ 7 \end{array}$$ _____

28.
$$\begin{array}{r} 9 \\ -\ 6 \end{array}$$ _____

29.
$$\begin{array}{r} 7 \\ -\ 3 \end{array}$$ _____

30.
$$\begin{array}{r} 8 \\ -\ 2 \end{array}$$ _____

31.
$$\begin{array}{r} 9 \\ -\ 3 \end{array}$$ _____

32.
$$\begin{array}{r} 9 \\ -\ 8 \end{array}$$ _____

33.
$$\begin{array}{r} 10 \\ -\ 3 \end{array}$$ _____

34. Math Journal Explain how to count up to find $11 - 9$. Why wouldn't you count back?

★ PROBLEM SOLVING

Use the clock to count back or count up. Solve each problem.

35. A play was 3 hours long. It ended at 6:00 P.M. At what time did it start?

_____

36. A movie began at 4:30 P.M. It was over at 6:30 P.M. How long was the movie?

_____

💡 Critical Thinking

Martha pays for the balloon with 2 quarters and a dime. How much change does she get? Explain how you can count up to make change.

57 ¢

_____

_____

Counting up is great for making change.

UN-FRAME IT!

A ten frame can help you learn and remember harder facts.

$13 - 5 = ?$

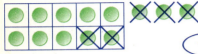

Think 13 = 10 + 3.

Subtracting 5 is the same as taking away 3 and then taking away 2 more.

$13 - 5 = 8$

$14 - 9 = ?$

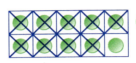

Think 14 − 9 = 14 − 10 plus 1 more.

Subtracting 9 is the same as taking away 10 and adding back 1.

$14 - 9 = 5$

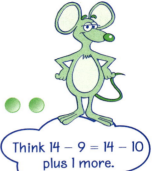

Your Turn

Draw lines to match the fact with the picture. Write the difference.

1.  $16 - 7 =$ _____

2.  $16 - 9 =$ _____

3. $15 - 8 =$ _____

Find each difference. Use a ten frame and counters if you like.

4. $13 - 6 =$ _____   5. $15 - 9 =$ _____   6. $17 - 8 =$ _____

7. $15 - 7 =$ _____   8. $11 - 5 =$ _____   9. $14 - 9 =$ _____

10. $19 - 9 =$ _____   11. $12 - 8 =$ _____   12. $16 - 7 =$ _____

Subtract.

13. 15
− 7

14. 13
− 6

15. 12
− 3

16. 14
− 6

17. 16
− 7

REMEMBER: Use counting up and counting back too!

18. 11
− 2

19. 16
− 9

20. 8
− 0

21. 10
− 7

22. 12
− 4

23. 15
− 8

24. 10
− 5

25. 5
− 1

26. 16
− 8

27. 14
− 7

28. 18
− 9

29. 6
− 3

PROBLEM SOLVING

30. Joan had 14 baseball cards. She lost 8 of them. How many cards does she still have?

_____

31. Kim had 15 stickers. She used 9 of them. Does she have enough left to put 4 on each of two pages of her sticker book? Explain.

_____

_____

_____

Critical Thinking

Write + or − in each circle to make the number sentence true.

1. 8 ◯ 6 ◯ 2 = 4

2. 13 ◯ 4 ◯ 6 = 3

3. 9 ◯ 6 ◯ 8 = 7

4. 17 ◯ 9 ◯ 3 = 11

5. 12 ◯ 5 ◯ 3 = 10

6. 14 ◯ 5 ◯ 4 = 5

ADD TO SUBTRACT?

If you can add, then you can subtract!

What if you bought this box of crayons? You open it up and there are only 8 crayons in the box. How can you find out how many are missing?

You can subtract.

12 − 8 = ?   (Think 8 + ? = 12.)

8 + 4 = 12   So 12 − 8 = 4.

These are **related facts.** They use the same numbers.

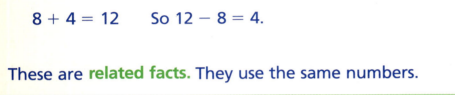 **Your Turn**

Add. Then use the addition fact to help you subtract.

1. 7 + 6 = _____

13 − 6 = _____

13 − 7 = _____

2. 6 + 4 = _____

10 − 4 = _____

10 − 6 = _____

3. 4 + 9 = _____

13 − 9 = _____

13 − 4 = _____

4. 5 + 6 = _____

11 − 6 = _____

11 − 5 = _____

5. 1 + 6 = _____

7 − 6 = _____

7 − 1 = _____

6. 8 + 2 = _____

10 − 2 = _____

10 − 8 = _____

7. 4 + 7 = _____

11 − 7 = _____

11 − 4 = _____

8. 6 + 8 = _____

14 − 8 = _____

14 − 6 = _____

9. 3 + 8 = _____

11 − 8 = _____

11 − 3 = _____

Add. Write the related subtraction facts.

TIP
Thinking of a related addition fact can help you subtract.

10. 4 + 5 = _____   **11.** 9 + 4 = _____

_____ − _____ = _____   _____ − _____ = _____

_____ − _____ = _____   _____ − _____ = _____

12. 7 + 3 = _____   **13.** 5 + 9 = _____   **14.** 8 + 1 = _____

_____ − _____ = _____   _____ − _____ = _____   _____ − _____ = _____

_____ − _____ = _____   _____ − _____ = _____   _____ − _____ = _____

Subtract. Use the strategies you've learned so far.

15.	**16.**	**17.**	**18.**	**19.**	**20.**
11	12	9	8	6	12
− 8	− 7	− 4	− 8	− 1	− 6

21.	**22.**	**23.**	**24.**	**25.**	**26.**
3	6	12	9	15	14
− 1	− 4	− 5	− 0	− 6	− 5

PROBLEM SOLVING

Will you add or subtract? Explain. Then solve.

27. Kevin bought 12 cookies. He ate 3. How many does he have left?

_____

_____

28. Mark made 8 kites. He bought 4 more. How many kites does he have now?

_____

_____

That's **AMAZING!**

Most 10-year-olds need 9 to 12 hours of sleep each night. Most adults need 7 to 9 hours. What is the difference between the least amount of sleep a child needs and the most an adult needs? How much sleep do you get each night?

_____

_____

IT'S ALL IN THE FAMILY!

The domino shows two addition facts.

Who's right? They both are!

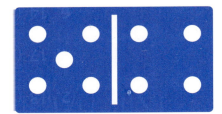

No! The domino shows two subtraction facts.

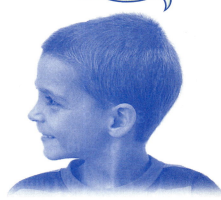

$$5 + 4 = 9 \qquad 9 - 5 = 4$$
$$4 + 5 = 9 \qquad 9 - 4 = 5$$

A group of related facts make up a **fact family**.

Knowing a fact family can also help you find a **missing addend**.

$8 + ? = 11$

Think $11 - 8 = 3$. 3 is the missing addend.

So $8 + 3 = 11$.

✏️ Your Turn

Write the fact family for each domino.

1.

____ + ____ = ____

____ + ____ = ____

____ − ____ = ____

____ − ____ = ____

2.

____ + ____ = ____

____ + ____ = ____

____ − ____ = ____

____ − ____ = ____

3.

____ + ____ = ____

____ + ____ = ____

____ − ____ = ____

____ − ____ = ____

Write the fact family for each group of numbers.

BEWARE!
Some fact families have only two members.

4. (8, 9, 17)

_____ + _____ = _____

_____ + _____ = _____

_____ − _____ = _____

_____ − _____ = _____

5. (9, 11, 2)

_____ + _____ = _____

_____ + _____ = _____

_____ − _____ = _____

_____ − _____ = _____

6. (4, 8, 4)

_____ + _____ = _____

_____ − _____ = _____

Find the missing addends.

7. $7 + \boxed{} = 15$

8. $\boxed{} + 9 = 16$

9. $6 + \boxed{} = 6$

10. $\boxed{} + 8 = 16$

11. $5 + \boxed{} = 13$

12. $4 + \boxed{} = 11$

Shade the oval of the fact that is not related to the others.

13.
Ⓐ $17 - 9 = 8$
Ⓑ $8 + 9 = 17$
Ⓒ $9 + 8 = 17$
Ⓓ $9 - 8 = 1$

14.
Ⓐ $8 + 5 = 13$
Ⓑ $8 - 5 = 3$
Ⓒ $5 + 8 = 13$
Ⓓ $13 - 8 = 5$

15.
Ⓐ $7 - 0 = 7$
Ⓑ $0 + 7 = 7$
Ⓒ $7 + 7 = 14$
Ⓓ $7 - 7 = 0$

★ PROBLEM SOLVING

Write a number sentence for each problem. Then solve.

16. Leon always loses his crayons. He had a box of 16. He has only 9 now. How many did he lose?

_____

17. Fran has 15 dolls. She has 8 on a shelf. The rest are in her closet. How many are in her closet?

_____

One For Fun!

FACT FAMILY RIDDLE

I am the least number in my fact family. The other two numbers are 7 and 9. What number am I? What are the facts in my family?

I am _____.

_____ + _____ = _____   _____ − _____ = _____

_____ + _____ = _____   _____ − _____ = _____

1. That's Just Ducky

There are 8 ducks in a pond. Some are white. Some are brown. There are 2 fewer white ducks than brown ducks. How many brown ducks are there?

2. Way to Pay

Kelly has 5 coins. She has 18¢. She wants to buy a banana for 8¢. Which of her coins can she use to pay?

3. Add or SUBTRACT

Write + or − in each circle to make the number sentence true.

1 ◯ 2 ◯ 3 ◯ 4 ◯ 5 = 9

4. GO THAT WAY!

Follow the arrows. Write the sum or difference in each circle.

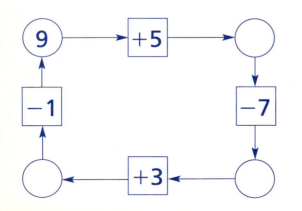

5. Family Time!

Ray says there are two fact families you can make with the numbers 5 and 8. Do you agree? Why or why not?

Pair Them Up!

Materials:
- 10 index cards for each player
- 1 marker, pen, or pencil for each player

Number of Players: 2 to 4

Goal: To collect the most Fact and Difference Cards

Game Rules

1 Each player makes 5 subtraction Fact Cards and 5 matching Difference Cards.

  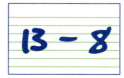 

Fact Card   Difference Card   Fact Card   Difference Card

2 Players put all their Fact Cards together in a pile. They put all their Difference Cards in another pile.

3 One player mixes the cards in each pile and places the cards face down.

4 The first player picks one card from each pile.

5 If the Fact and Difference Cards form a pair, the player keeps them both. If not, the player mixes the cards back into each pile.

6 Players take turns until all the pairs have been formed.

7 The player with the greatest number of pairs wins.

SPECIAL SUMS

Do you walk a lot in school? David does. He walks
300 feet from his classroom to the media center.
Then he walks 500 feet to the office.

How many feet does he walk altogether?

$300 + 500 = ?$

You can use facts you know and **patterns** to help you add.

Think $3 + 5 = 8$.

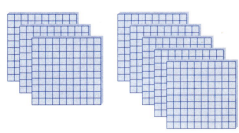

3 ones + 5 ones
$3 + 5 = 8$

3 tens + 5 tens
$30 + 50 = 80$

3 hundreds + 5 hundreds
$300 + 500 = 800$

Altogether, David walks 800 feet.

Your Turn

Look for patterns. Find each missing number.

1. $6 + 3 = $ _____

$60 + 30 = $ _____

$600 + 300 = $ _____

2. $4 + 9 = $ _____

$40 + 90 = $ _____

$400 + 900 = $ _____

3. $7 + 3 = $ _____

$70 + 30 = $ _____

$700 + 300 = $ _____

4. $7 + $ _____ $ = 15$

$70 + $ _____ $ = 150$

$700 + $ _____ $ = 1,500$

5. _____ $ + 8 = 13$

_____ $ + 80 = 130$

_____ $ + 800 = 1,300$

6. $9 + $ _____ $ = 16$

$90 + 70 = $ _____

_____ $ + 700 = 1,600$

Add.

> **REMEMBER:**
> Use facts you know to help you add tens and hundreds.

7. 40 + 30 = _____

8. 400 + 300 = _____

9. 80 + 20 = _____

10. 900 + 800 = _____

11. 500 + 500 = _____

12. 60
  + 80

13. 400
  + 800

14. 900
  + 700

15. 80
  + 70

16. 70
  + 60

Fill in the missing numbers in this addition table.

17.

+	100	200	400		700	900
200	300			700		1,100
	400			800	1,000	

★ **PROBLEM SOLVING**

The pictograph shows how many books Stacy, Jeff, Raoul, and Shanta read. Use the graph to answer each question.

Books Read

Stacy	📖 📖 📖 📖 📖
Jeff	📖 📖
Raoul	📖 📖 📖 📖
Shanta	📖 📖 📖

KEY: 📖 = 10 books

18. How many books did Stacy and Shanta read? _____

19. How many books did all four students read? _____

💡 **Critical Thinking**

> It's easy to add 10 or 100 to any number. Just find a pattern.

38 + 10 = 48	467 + 100 = 567
48 + 10 = 58	567 + 100 = 667
58 + 10 = 68	667 + 100 = 767

Look for patterns. Write a rule for adding 10 or 100 to any number.

_____

_____

_____

DOWN AND OVER

Did you know you can count on to add greater numbers too?

34 + 23 = ?

> Think 23 is the same as 2 tens and 3 ones.

Start at 34. First count on 2 tens. Go down 2 rows.

> Think 34.
> Count on 44, 54.

Now count on 3 ones. Go over 3 columns.

> Think 54.
> Count on 55, 56, 57.

So 34 + 23 = 57.

TIP
A hundred chart can help you add.

1	2	3	4	5	6	7	8	9	10
11	12	13	14	15	16	17	18	19	20
21	22	23	24	25	26	27	28	29	30
31	32	33	34	35	36	37	38	39	40
41	42	43	44	45	46	47	48	49	50
51	52	53	54	55	56	57	58	59	60
61	62	63	64	65	66	67	68	69	70
71	72	73	74	75	76	77	78	79	80
81	82	83	84	85	86	87	88	89	90
91	92	93	94	95	96	97	98	99	100

Your Turn

Count on to find each sum.

1. 54
    + 20

2. 30
    + 37

3. 48
    + 20

4. 62
    + 3

5.   2
    + 49

6. 74
    + 20

7. 54
    + 2

8. 17
    + 10

9. 97
    + 1

10. 69
    + 3

11. 24
    + 3

12. 30
    + 61

Add. Use a hundred chart if you like.

13. 57 + 33 = _____

14. 14 + 50 = _____

15. 26 + 10 = _____

16. 51 + 20 = _____

17. 3 + 42 = _____

18. 30 + 42 = _____

19. 66 + 21 = _____

20. 38 + 32 = _____

21. 45 + 2 = _____

22. 19 + 50 = _____

23. 26 + 62 = _____

★ PROBLEM SOLVING

Help Morty through the mouse maze. It's really a hundred chart, but don't tell Morty! Use the hundred chart on page 24 to solve each problem.

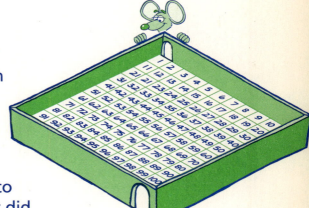

24. Morty dropped some cheese on a number. Then he ran down 2 rows in the same column. He stopped on 67. On which number did he drop the cheese? _____

25. From 67, Morty ran 3 rows down and 2 spaces to the left. There he fell asleep. On which number did he sleep? _____

26. When Morty wakes up, how many spaces will he have to go to get to 100? _____

One For Fun!

MOUSE HOUSE

Morty loves his maze. He wants to move in. Write a funny addition story about Morty's new house. (Do not write the sum!) Have a friend find the sum using a hundred chart.

_____

_____

_____

GREAT GUESSES

Pam loves to read. She read 37 books last year and 52 this year. About how many books did she read?

37 + 52 is about ?

When you don't need an exact answer, you can **estimate**.

One way to estimate is to round each addend to the nearest 10. Then add.

37 — rounds to →	40	37 is close to 40
+ 52 — rounds to →	+ 50	52 is close to 50
	90	

So 37 + 52 is about 90.

You can estimate with greater numbers too.

268 + 126 is about ?

Now round to the nearest 100. Then add.

268 — rounds to →	300	268 is close to 300
+ 126 — rounds to →	+ 100	126 is close to 100
	400	

So 268 + 126 is about 400.

TIP
Some numbers, like 25, are halfway between 2 tens. In that case, round up. Round 25 up to 30.

Your Turn

Round to the nearest 10. The first one is done for you.

1. 58 __60__ 2. 17 _____ 3. 44 _____ 4. 52 _____ 5. 35 _____

6. 73 _____ 7. 54 _____ 8. 87 _____ 9. 65 _____ 10. 61 _____

Round to the nearest 100.

11. 132 _____ 12. 289 _____ 13. 511 _____ 14. 479 _____ 15. 738 _____

16. 319 _____ 17. 571 _____ 18. 408 _____ 19. 665 _____ 20. 450 _____

Round each addend. Add to estimate the sum. The first one is done for you.

21. 22 + 39

$$\underline{20} + \underline{40} = \underline{60}$$

22. 36 + 58

_____ + _____ = _____

23. 27 + 64

_____ + _____ = _____

24. 78 + 25

_____ + _____ = _____

25. 308 + 188

_____ + _____ = _____

26. 567 + 78

_____ + _____ = _____

27. 196 + 725

_____ + _____ = _____

28. 431 + 399

_____ + _____ = _____

29. 651 + 234

_____ + _____ = _____

Estimate. Write < or > in each circle.

30. 55 + 37 ◯ 80

31. 126 + 30 ◯ 200

32. 288 + 160 ◯ 300

33. Math Journal Megan says, "I can make a closer estimate if I round only one addend than if I round them both." Explain what she means.

Problem Solving

Use the graph to answer each question.

34. About how many tickets did Andy and Carla sell together? _____

35. About how many tickets did Mika and Sam sell together? _____

Tickets Sold

Critical Thinking

Pretend you are in a restaurant. You want to make sure you have enough money to pay for your meal. You look at the prices and decide to estimate. Should you round all the prices up or down? Explain.

_____

_____

_____

GET IT TOGETHER!

In the last lesson you learned how to estimate. What if you need an exact answer? How can you find $35 + 28$?

Try using base-ten models.

$35 + 28 = ?$

Show each number with models.

Put the models together.

Regroup 10 ones as 1 ten if you can.

$35 + 28 = 63$

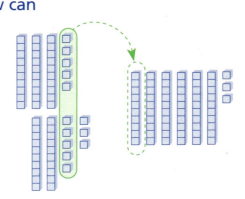

Your Turn

Decide if you need to regroup. Circle 10 ones if you can. Then add.

1. 42
    + 25

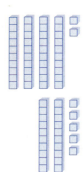

2. 27
    + 36

3. 45
    + 23

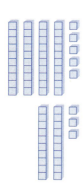

4. 64
    + 18

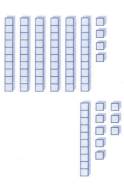

Add. Use base-ten models.

5. 56 + 25 = _____

6. 43 + 31 = _____

7. 39 + 45 = _____

8. 9 + 70 = _____

9. 82 + 12 = _____

10. 76 + 8 = _____

11. 34
    + 48

12. 52
    + 4

13. 29
    + 63

14. 45
    + 19

15. 57
    + 31

16. 12
    + 72

17. 31
    + 15

18. 46
    + 53

19.   3
    + 29

20. 24
    + 47

21. 89
    + 11

22. 42
    + 49

23. **Math Journal** How do you know when you need to regroup? Explain. Give an example.

Problem Solving

The chart shows the books sold at the school book fair. Estimate or use models to solve.

24. On which day were more books sold? _____

25. Should the school order more fiction books or more nonfiction books to sell next year? Explain. _____

_____

Books Sold		
Day	Fiction	Nonfiction
1	38	29
2	23	46

Critical Thinking

TIP
Dimes and pennies are just like tens and ones.

Fill in the blanks. Then write the sum. (Hint: Regroup 10 pennies for 1 dime.)

1.   14¢
    + 23¢

_____ dimes _____ pennies

_____ ¢

2.   33¢
    + 27¢

_____ dimes _____ pennies

_____ ¢

GET IN STEP!

There are 28 children in Mrs. Brown's class and 34 children in Mr. Ortez's class. Each child needs a book. How many books are needed?

$$28 + 34 = ?$$

You can follow these steps to add. Or you can use base-ten models.

Step 1

Add the ones.
Regroup if you can.

8 ones + 4 ones = 12 ones
Regroup 12 ones as 1 ten and 2 ones.

```
  1
  28
+ 34
───
   2
```

TIP
Estimate before or after you add. Then use your estimate to see if your answer is reasonable.

Step 2

Add the tens.
Regroup if you can.

1 ten + 2 tens + 3 tens = 6 tens

```
  1
  28
+ 34
───
  62
```

62 books are needed.

Your Turn

Estimate. Then find each sum. The first estimate is done for you.

1.
```
  16      20
+ 53   + 50
        ───
         70
```

2.
```
  50¢
+ 14¢
```

3.
```
  19¢
+ 39¢
```

4.
```
  22
+ 68
```

5.
```
  28¢
+ 48¢
```

6.
```
  37
+ 32
```

7.
```
  45¢
+ 35¢
```

8.
```
  68
+ 27
```

9.
```
  72
+ 18
```

10.
```
  54
+  9
```

11.
```
  65
+ 15
```

12.
```
  77¢
+ 16¢
```

Add. Rewrite the exercise in vertical form. Estimate to see if your answer is reasonable.

13. 43 + 50

14. 39¢ + 42¢

15. 64 + 8

16. 17¢ + 59¢

17. 77 + 13

18. 9¢ + 58¢

19. 70 + 23

20. 69¢ + 21¢

21. 24 + 67

22. 32 + 29

23. 52 + 41

24. 66 + 14

Problem Solving

25. Brad bought a pencil and a marker. How much did he spend? _____

26. Rosa spent 53¢. Which two items did she buy?

_____

SCHOOL SALE

MARKER . . . . 45¢
SCISSORS . . . 35¢
ERASER . . . . 18¢
PENCIL . . . . . 13¢

 ## Critical Thinking

Look at how Senja added. Does her answer make sense? What did she do wrong? Explain.

$$\begin{array}{r} 36 \\ + 59 \\ \hline 815 \end{array}$$

_____

_____

_____

GET IN STEP AGAIN!

Adding 3-digit numbers is like adding 2-digit numbers with an extra step. You have to add hundreds too.

Follow these steps to find 256 + 167.

You can estimate first. 300 + 200 = 500. The sum should be close to 500.

Step 1

Add the ones.
Regroup if you can.

6 ones + 7 ones = 13 ones
Regroup 13 ones as 1 ten and 3 ones.

$$\begin{array}{r} 1 \\ 256 \\ + 167 \\ \hline 3 \end{array}$$

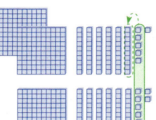

Step 2

Add the tens.
Regroup if you can.

1 ten + 5 tens + 6 tens = 12 tens
Regroup 12 tens as 1 hundred and 2 tens.

$$\begin{array}{r} 11 \\ 256 \\ + 167 \\ \hline 23 \end{array}$$

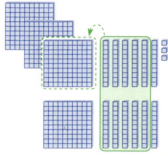

Step 3

Add the hundreds.
Regroup if you can.

1 hundred + 2 hundreds + 1 hundred = 4 hundreds

$$\begin{array}{r} 11 \\ 256 \\ + 167 \\ \hline 423 \end{array}$$

256 + 167 = 423

✏️ Your Turn

Estimate. Then find each sum. The first estimate is done for you.

1. $\begin{array}{r} 276 \\ + 319 \\ \hline \end{array}$ $\begin{array}{r} 300 \\ + 300 \\ \hline 600 \end{array}$

2. $\begin{array}{r} 462 \\ + 85 \\ \hline \end{array}$

3. $\begin{array}{r} 129 \\ + 750 \\ \hline \end{array}$

4. $\begin{array}{r} 591 \\ + 62 \\ \hline \end{array}$

5. $\begin{array}{r} 622 \\ + 198 \\ \hline \end{array}$

6. $\begin{array}{r} 426 \\ + 383 \\ \hline \end{array}$

7. $\begin{array}{r} 625 \\ + 82 \\ \hline \end{array}$

8. $\begin{array}{r} 53 \\ + 749 \\ \hline \end{array}$

Add. Estimate to see if your answer is reasonable.

BEWARE!
When adding 3-digit numbers, you may have to regroup once, twice, three times, or not at all.

9. 378
    + 215

10. 437
    + 56

11. 806
    + 98

12. 457
    + 123

13. 555
    + 55

14. 506
    + 179

15. 24
    + 333

16. 496
    + 7

17. 228
    + 167

18. 47
    + 891

19. 704
    + 148

20. **Math Journal** Kalita says, "I like to use play money to help me add 3-digit numbers. I use 1-, 10-, and 100-dollar bills." Explain how Kalita can use money to add.

PROBLEM SOLVING

Use the map to solve each problem.

21. How far is it from Joplin to St. Louis driving only through Springfield?

    _____ miles

22. How far is it from Poplar Bluff to Jefferson City going through St. Louis?

    _____ miles

MISSOURI

Jefferson City   St. Louis
           129
137
     224
                   156
Springfield
76   192
Joplin   Poplar
                    Bluff

Critical Thinking

Look at how Gary added. What did he do wrong? Explain.

      507
    + 136
      633

AS A MATTER OF FACT, IT'S EXACT!

Carlos wants to buy a bag of chips and an apple. He wants to know exactly how much that will cost. He doesn't have a paper and pencil. How can he find out?

Think 59¢ + 27¢ = ?

SNACK SALE
Chips 59¢ a bag
Apples 27¢ each
Drinks 62¢ each

Here are some strategies he can use.

He can break each number apart to make it easier to add.

$59 = 50 + 9$

$27 = 20 + 7$

Add the tens. $50 + 20 = 70$
Add the ones. $9 + 7 = 16$

Add the tens and ones. $70 + 16 = 86$

He can round to get numbers that are easier to work with.

Round 59 up to 60.
This is the same as adding 1.

Now add. $60 + 27 = 87$

Subtract the 1 you added.
$87 - 1 = 86$

It will cost 86¢ for a bag of chips and an apple.

Your Turn

Show how to break each number apart.

1. 48 = _____ + _____

2. 62 = _____ + _____

3. 19 = _____ + _____

4. 87 = _____ + _____

5. 36 = _____ + _____

6. 98 = _____ + _____

7. 108 = _____ + _____

8. 299 = _____ + _____

9. 597 = _____ + _____

10. 113 = _____ + _____

11. 130 = _____ + _____

12. 999 = _____ + _____

13. 51 = _____ + _____

14. 37 = _____ + _____

15. 416 = _____ + _____

Add mentally. Break numbers apart or round to get easier numbers. Looking back at Exercises 1–15 may help you.

REMEMBER:
Sometimes you can just count on to add mentally.

16. 48 + 20 = _____

17. 37 + 62 = _____

18. 48 + 19 = _____

19. 3 + 89 = _____

20. 22 + 87 = _____

21. 36 + 51 = _____

22. 108 + 62 = _____

23. 416 + 20 = _____

24. 130 + 999 = _____

25. 597 + 299 = _____

26. 299 + 36 = _____

27. Math Journal   Explain one way to find 79 + 43 using mental math.

Problem Solving

Do these in your head.

28. If 285 + 179 = 464,

then 685 + 179 = _____

29. If 205 + 54 = 259,

then 405 + 354 = _____

30. If 166 + 235 = 401,

then 466 + 235 = _____

31. If 333 + 222 = 555,

then 533 + 422 = _____

Critical Thinking

Write *always*, *sometimes*, or *never* to make each sentence true.

1. A rounded number _____ has a zero in the ones place.

2. The sum of two 2-digit numbers is _____ a 4-digit number.

3. You _____ round addends to the nearest hundred to estimate.

FOUR STEPS FOR FOUR DIGITS

TIP
Adding 4-digit numbers is like adding 3-digit numbers but with an extra step. You need to add the thousands.

One month the third graders collected 2,809 cans to recycle. The next month they collected 1,738 cans. How many cans did they collect?

$$2,809 + 1,738 = ?$$

Estimate. $3,000 + 2,000 = 5,000$ The sum should be *about* 5,000.

Step 1	**Step 2**	**Step 3**	**Step 4**
Add the ones. Regroup if you can.	Add the tens. Regroup if you can.	Add the hundreds. Regroup if you can.	Add the thousands. Regroup if you can.
$\begin{array}{r} 1 \\ 2,809 \\ + 1,738 \\ \hline 7 \end{array}$	$\begin{array}{r} 1 \\ 2,809 \\ + 1,738 \\ \hline 47 \end{array}$	$\begin{array}{r} 1\ \ 1 \\ 2,809 \\ + 1,738 \\ \hline 547 \end{array}$	$\begin{array}{r} 1\ \ 1 \\ 2,809 \\ + 1,738 \\ \hline 4,547 \end{array}$

The third graders collected 4,547 cans.

To check your work, compare the sum to the estimate.
4,547 is about 5,000 so the answer makes sense.

Your Turn

Estimate mentally. Then find the sum.

1. 3,605 + 333	2. 5,605 + 3,214	3. 5,456 + 1,321	4. 5,020 + 198
5. 5,158 + 1,126	6. 2,782 + 29	7. 932 + 2,075	8. 2,358 + 1,747
9. 6,054 + 1,980	10. 3,698 + 701	11. 7,373 + 2,442	12. 4,163 + 2,357
13. 1,047 + 499	14. 3,591 + 2,325	15. 7,329 + 2,577	16. 4,282 + 4,822

Estimate mentally. Rewrite the exercise in vertical form. Then add.

17. 4,815 + 1,705

18. 385 + 3,041

19. 47 + 1,006

20. 5,520 + 1,609

21. 5,378 + 123

22. 4,030 + 97

23. 784 + 2,340

24. 6,058 + 1,942

25. 7,727 + 486

Shade the oval for the correct answer.

26. 1,158
      +   935

Ⓐ 1,093
Ⓑ 1,083
Ⓒ 2,083
Ⓓ 2,093

27. 1,617
      +   283

Ⓐ 1,434
Ⓑ 1,890
Ⓒ 1,900
Ⓓ 1,910

28. 4,321 + 375

Ⓐ 4,696
Ⓑ 4,796
Ⓒ 7,071
Ⓓ 8,071

★ PROBLEM SOLVING

Megan used a calculator to input 4,815 + 4,750. She pressed  and the sum 5,290 appeared. "That answer is not reasonable!" Megan said.

29. Why isn't 5,290 reasonable?

_____

_____

_____

30. What might Megan have done wrong?
(Hint: Look at the digits in each number.)

_____

_____

_____

That's AMAZING!

The Akashi-Ohashi Bridge in Japan has the longest span in the world. It spans 5,840 feet. How far would you have to walk to cross the bridge and come back?

_____

_____

ADDING DOWN, CHECKING UP

Mrs. Brown's class collects items for recycling. How many items did they collect?

223 + 307 + 92 = ?

Follow the same steps you used to add two numbers.

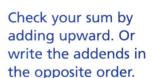

Items Collected for Recycling	
cans	223
newspapers	307
bottles	92

Step 1

Add the ones. Regroup if you can.

```
  1
 223
 307
+ 92
───────
   2
```

Step 2

Add the tens. Regroup if you can.

```
 1 1
 223
 307
+ 92
───────
  22
```

Step 3

Add the hundreds. Regroup if you can.

```
 1 1
 223
 307
+ 92
───────
 622
```

Check your sum by adding upward. Or write the addends in the opposite order.

```
 1 1
223    92
307   307
+ 92  + 223
───────────
       622
```

The class collected 622 items for recycling.

TIP

A calculator can be a big help when adding greater numbers. If you can, use one to check your work.

Your Turn

Add. Then change the order of the addends and add again.

1.
```
  28
  51
  26
+ 12
```

2.
```
 385
  49
+ 206
```

3.
```
 504
  68
 175
+  8
```

4.
```
 367
  34
 204
+ 59
```

5.
```
 2,178
  405
   76
+   8
```

6.
```
 304
 678
 555
+ 540
```

7.
```
 1,872
  390
  504
+  13
```

Rewrite each exercise in vertical form. Then add.

8. 209 + 290 + 902

9. 815 + 86 + 607 + 8

10. 563 + 92 + 4,306

11. 349 + 568 + 801

12. 54 + 543 + 5,432

13. 123 + 56 + 789

★ Problem Solving

14. Show how to put these blocks into two stacks so that each stack has the same sum. Write the number in each box. (Hint: You may want to round the numbers to estimate where the blocks belong.)

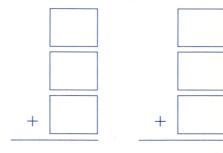

One for Fun!

HALF A TON OF FUN

Use a calculator to solve this one!

Elevator Capacity:
$\frac{1}{2}$ ton or
1,000 pounds

About how many adults can ride in this elevator at one time?

Ask a few adults how much they weigh. If they won't tell you, make up some reasonable weights! Find the sum of the weights.

_____

WHAT'S THE TOTAL?

How much would a slice of pizza and a milk shake cost?

$2.88 + $1.35 = ?

Adding money is like adding whole numbers.
But don't forget to write dollar signs and decimal points.

Estimate. $3.00 + $1.00 = $4.00
The sum should be *about* $4.00.

Tony's Pizzeria
Special Today:

Pizza $2.88 each slice
Milk shake $1.35 each

You can use a calculator to add money. Be sure to press ⊡ when you need to.

Step 1
Write each amount. Line up the cents and the dollars.

$2.88
+ 1.35

Step 2
Add the way you would add whole numbers.

^{1 1}
$2.88
+ 1.35
4 23

Step 3
Write a dollar sign and a decimal point in the sum.

$2.88
+ 1.35
$4.23

A slice of pizza and a milk shake would cost $4.23.

✎ Your Turn

Estimate mentally. Then add. Don't forget to write a decimal point and a dollar sign in the sum.

BEWARE!
Dollar signs do not show up on calculator displays. You must write them in the sums.

1. $1.29
    + 3.54

2. $4.53
    + 0.27

3. $0.75
    + 0.19

4. $3.24
    + 3.69

5. $1.50
    + 0.38

6. $0.69
    + 0.75

7. $0.38
    + 0.95

8. $9.50
    + 4.78

9. $2.25
    + 0.95

10. $3.98
    + 2.75

11. $4.67
    + 1.85

12. $8.71
    + 0.75

13. $9.38
    + 6.14

14. $7.12
    + 4.85

Find each sum.

15. $6.00 + $3.00 = _____

16. $0.40 + $0.25 = _____

17. $0.30 + $1.50 = _____

18. $7.30
    + 2.50
    _____

19. $1.25
    + 6.70
    _____

20. $3.45
    + 2.34
    _____

21. $0.63
    + 0.25
    _____

22. $5.63
    + 4.25
    _____

23. $4.50
      0.67
    + 3.00
    _____

24. $0.79
    + 0.53
    _____

25. $0.65
      2.88
      1.99
    + 0.07
    _____

★ Problem Solving

26. Kayla wants to order a pair of hiking boots and a pair of socks. What will her total be, including the delivery charge?

    _____

27. Kayla has a $50 gift certificate for use at Comp Camp. Which items can she buy for $50?

    _____

    _____

Comp Camp

hiking boots	$39
socks	$6
camp stove	$23

Delivery Charge:
$2 per item

Critical Thinking

Gina and Marla each need $1.50 for the bus. Which of them could lend the other money so they both can ride? How much would she have to lend?

Gina has

Marla has

_____ could lend _____ $ _____.
   (name)   (name)   (How much?)

QUICK TOTALS

Kareem is in a rush. He buys these items. He wanted a very quick estimate. He adds only the front-end digit of each price. What is his estimate?

Soap powder	$4.79 ⟶	$4.00
Cereal	3.10 ⟶	3.00
Ground beef	4.23 ⟶	4.00
Lettuce	1.29 ⟶	1.00
		$12.00

TIP
To use front-end estimation, add the front digits. Make believe the other digits are zeros.

Front-end digits are the ones that come first in a number. Use **front-end estimation** when most of the addends have the same number of digits.

Kareem's very quick estimate is $12.

Your Turn

Use front-end estimation. Write the estimate.
The first one is started for you.

BEWARE!
Add only those front digits that have the same place value.

1. $52 $50
    + 4   + 0

2. 475
    + 573

3. 76
    + 27

4. 5,139
    + 1,580

5. 2,005
    + 139

6. 607
    + 297

7. 847
    + 47

8. 227
    376
    + 519

9. $78
    16
    + 81

10. 3,055
    1,897
    + 2,243

11. 129
    82
    + 305

12. 38
    46
    67
    + 51

13. 450
    326
    147
    + 62

14. $17.95
    3.10
    15.50
    + 32.50

15. $3.80
    2.79
    0.55
    + 1.37

Use front-end estimation to find a very quick estimate.

16. $2.30	17. $0.35	18. $0.75	19. $1.22
0.25	0.78	2.66	4.58
1.17	0.95	0.89	0.39
+ 3.08	+ 0.45	+ 1.29	+ 2.47

20. Look back at Exercise 18. Use rounding to estimate again. What is your

new estimate? Is it less than or greater than your first estimate? _____

21. Math Journal Why are the front-end digits the most important ones to use for estimating?

⭐ **PROBLEM SOLVING**

22. What is a very quick estimate of the total cost of these items? Use front-end estimation.

_____

23. How could you use the other digits and your front-end estimate to make a closer estimate?

_____

24. Why might you want to make an estimate before buying something?

_____

$12.95

$1.59

$22.95

Critical Thinking

Look at each estimate. Tell why a front-end estimate is usually a *low estimate*.

_____

_____

	Rounded Estimate	Front-End Estimate
462 →	500	400
72 →	100	0
138 →	100	100
+ 315 →	+ 300	+ 300
	1,000	800

43

1. SOME SUMS!

Add across. Add down.

240	+	263	=	
+		+		+
162	+	192	=	
=		=		=
	+		=	

2. Number Pals

1,441 is a **palindrome**. Read it forward and backward. It's the same. Find your own palindrome. Here's how.

1. Write a 2-digit number.

2. Below it, write the number backward.

3. Add. (Is your sum a palindrome?)

4. If not, write the digits of the sum backward.

5. Keep adding and writing the sums backward. Soon you'll have your palindrome!

3. Number Path

Use these digits to make the greatest 4-digit number possible. Use them to make the least 4-digit number. Add the numbers. Write the sum.

4. Decode It!

Find the digit that stands for each letter.

$$\begin{array}{r} ABC \\ + ABC \\ \hline 718 \end{array}$$

A = _____ B = _____ C = _____

5. Pair Up!

Look at the numbers in each box. Write two of them to match each sum.

931   737   409

239   765

1. 1,696 = _____ + _____

2. 1,146 = _____ + _____

3. 1,004 = _____ + _____

4. 648 = _____ + _____

Spin a Sum!

Number of Players: 2

Goal: To be the first player to get 5 points

Game Rules

Materials:
- one spinner divided into 8 sections labeled 1, 2, 3, 4, 5, 6, 7, 8
- paper and pencil for each player

1. Spin the spinner 3 times. Record the digit you spin each time. Use the digits to write the greatest 3-digit number possible. Now do all this again! Write your two numbers as an addition example.

2. The other player does the same thing.

3. Players each find the sum of their own 3-digit numbers.

4. The player with the greater sum gets 1 point.

5. Players continue taking turns trying to make the greater sum.

6. The first player to get 5 points wins the game.

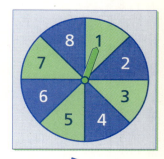

$$\begin{array}{r} 421 \\ + \ 831 \\ \hline \end{array}$$

$$\begin{array}{r} 221 \\ + \ 833 \\ \hline \end{array}$$

SPECIAL DIFFERENCES

Spencer ran 400 feet. Shawn ran 200 feet. How much farther did Spencer run?

$400 - 200 = ?$

You can use facts you know and patterns to help you subtract.

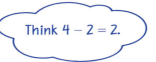

Think $4 - 2 = 2$.

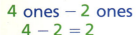

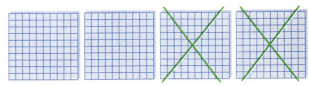

4 ones − 2 ones
$4 - 2 = 2$

4 tens − 2 tens
$40 - 20 = 20$

4 hundreds − 2 hundreds
$400 - 200 = 200$

Spencer ran 200 feet farther.

Your Turn

Look for patterns. Find the missing numbers.

1. $7 - 4 = $ _____

$70 - 40 = $ _____

$700 - 400 = $ _____

2. $13 - 9 = $ _____

$130 - 90 = $ _____

$1,300 - 900 = $ _____

3. $17 - 8 = $ _____

$170 - 80 = $ _____

$1,700 - 800 = $ _____

4. $9 - $ _____ $= 6$

$90 - $ _____ $= 60$

$900 - $ _____ $= 600$

5. _____ $- 6 = 9$

_____ $- 60 = $ _____

_____ $- 600 = $ _____

6. $11 - 4 = $ _____

_____ $- 40 = 70$

$1,100 - $ _____ $= 700$

7. $15 - 6 = $ _____

_____ $- 60 = 90$

$1,500 - 600 = $ _____

8. $12 - 8 = $ _____

$120 - $ _____ $= 40$

_____ $- 800 = 400$

9. $13 - 5 = $ _____

$130 - 50 = $ _____

_____ $- 500 = 800$

Subtract.

TIP
Say each exercise to yourself as you do it. It helps!

10. 50 − 20 = _____

11. 90 − 30 = _____

12. 120 − 90 = _____

13. 800 − 500 = _____

14. 1,500 − 700 = _____

15. 1,000 − 400 = _____

16. 900 − 400 = _____

17. 1,100 − 200 = _____

18.
700
− 300

19.
1,400
− 600

20.
1,500
− 800

21.
1,300
− 900

22.
6,000
− 3,000

PROBLEM SOLVING

Cara's family is driving from Austin to Amarillo. They have already driven 50 miles.

23. About how much farther do they have to go to get to Big Spring? _____

24. About how far are they now from Amarillo? _____

25. Why might you want to know about how much farther there is to go on a trip?

_____

Critical Thinking

62 − 10 = 52   735 − 100 = 635

52 − 10 = 42   635 − 100 = 535

42 − 10 = 32   535 − 100 = 435

Look for patterns. Write a rule for subtracting 10 or 100 from any number.

_____

_____

_____

It's easy to subtract 10 or 100 from any number. Just look for patterns.

47

UP AND OVER

You can count back to subtract one 2-digit number from another.

$72 - 33 = ?$

> Think 33 is the same as 3 tens and 3 ones.

Start at 72.
Count back 3 tens.

> Think 72.
> Count back
> 62, 52, 42.

Now count back 3 ones.

> Think 42.
> Count back
> 41, 40, 39.

So $72 - 33 = 39$.

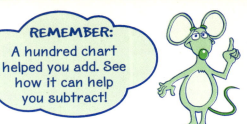

> **REMEMBER:**
> A hundred chart helped you add. See how it can help you subtract!

1	2	3	4	5	6	7	8	9	10
11	12	13	14	15	16	17	18	19	20
21	22	23	24	25	26	27	28	29	30
31	32	33	34	35	36	37	38	39	40
41	42	43	44	45	46	47	48	49	50
51	52	53	54	55	56	57	58	59	60
61	62	63	64	65	66	67	68	69	70
71	72	73	74	75	76	77	78	79	80
81	82	83	84	85	86	87	88	89	90
91	92	93	94	95	96	97	98	99	100

Your Turn

Count back to find each difference.

1.	2.	3.	4.	5.	6.
64 − 30	48 − 20	97 − 3	22 − 12	41 − 32	58 − 31

7.	8.	9.	10.	11.	12.
82 − 13	65 − 22	55 − 20	31 − 21	73 − 2	89 − 11

13.	14.	15.	16.	17.	18.
94 − 12	46 − 19	78 − 56	60 − 47	98 − 81	66 − 35

Use a hundred chart to subtract.

REMEMBER:
First count back by tens, then by ones.

19. 47 − 20 = _____

20. 34 − 17 = _____

21. 56 − 39 = _____

22. 63 − 25 = _____

23. 57 − 18 = _____

24. 70 − 28 = _____

25. 85 − 37 = _____

26. 68 − 27 = _____

27. 55 − 16 = _____

28. 74 − 38 = _____

29. 80 − 48 = _____

30. Math Journal Tell why a hundred chart makes it easy to subtract.

PROBLEM SOLVING

Remember Morty the mouse from Lesson 10? He's the mouse with a house like a hundred chart.

31. Start at 1. Find the seventh row of Morty's house. Now look for a space with two of the same digits. Morty ate dinner on a space that is 3 rows up from here in the same column. On which space did he eat dinner?

_____

32. See Morty on space 75? He plans to meet a mouse friend 4 rows up in the same column and 2 spaces to the right. On which space will the mice meet?

_____

One For Fun!

IT'S A SECRET!

Follow the clues to find the secret number.

- It is greater than the sum of 37 and 23.
- It is less than the difference between 94 and 31.
- It is odd.

The secret number is _____.

"GIANT" ESTIMATES

Amy thinks her dad is a giant. He is 6 feet 1 inch tall. That's the same
as 73 inches. Amy is just 49 inches tall. About how much taller
is Amy's dad than Amy?

73 − 49 is about?

You can estimate by rounding each number
to the nearest 10.

73 − 49 is about 20.

73	— rounds to →	70
− 49	— rounds to →	− 50
		20

Her dad is about 20 inches taller than Amy.

A sign at the zoo says that the giraffe is 228 inches
tall. About how much taller is the giraffe than Amy's dad?

228 − 73 is about?

Now round to the nearest 100.

228 − 73 is about 100.

228	— rounds to →	200
− 73	— rounds to →	− 100
		100

The giraffe is about 100 inches taller than Dad.

✎ Your Turn

Round to the nearest 10. Subtract to estimate the difference.
The first one is done for you.

1. 68 70
    − 29 − 30
    _____
         40

2. 73
    − 25

3. 162
    − 97

4. 392
    − 89

Round to the nearest 100. Subtract to estimate the difference.

5. 612
    − 395

6. 832
    − 307

7. 709
    − 693

8. 905
    − 325

Estimate. Write < or > in each circle.

9. 72 − 43 ◯ 20

10. 57 − 12 ◯ 30

11. 85 − 24 ◯ 40

12. 101 − 69 ◯ 50

13. 248 − 152 ◯ 84

14. 195 − 88 ◯ 200

Estimate each difference in your head. Write the estimate.

15. 92 − 43 _____

16. 521 − 384 _____

17. $6.07 − $3.86 _____

18. 817 − 312 _____

19. $5.12 − $4.97 _____

20. 923 − 117 _____

21. $735 − $289 _____

22. 878 − 427 _____

23. 614 − 185 _____

24. Math Journal Mario says, "Rounding both numbers down to estimate is the same as using front-end estimation." Tell whether or not Mario is correct. Give some examples.

★ **PROBLEM SOLVING**

Use the chart to find each estimate.

25. About how much taller is the Empire State Building than the Woolworth Building?

_____

26. About how much taller is the World Trade Center than the RCA Building?

_____

Tall Buildings in New York City

Building	Height (in feet)
World Trade Center	1,377
Empire State Building	1,250
Citicorp Building	915
RCA Building	850
Woolworth Building	792
General Motors Building	705

Critical Thinking

Which is the better estimate for each difference? Explain.

1. 853 − 532

a little more than 300

a little less than 300

_____

_____

2. $6.67 − $4.93

a little more than $2.00

a little less than $2.00

_____

_____

TAKE IT APART!

Now you know how to estimate differences.
But do you know how to find an exact difference?
Use base-ten models for help.

$$65 - 38 = ?$$

Show 65.
Take away 38.
That's 3 tens and 8 ones.
But there are only 5 ones!
Regroup 1 ten as 10 ones.

TIP
When you add, regroup ones to get more tens. When you subtract, regroup tens to get more ones.

Now you can subtract.

$$65 - 38 = 27$$

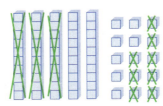

Your Turn

Subtract. Cross out 1 ten and draw 10 ones if you need to regroup.

1. 56
   − 29
   _____

2. 83
   − 45
   _____

3. 78
   − 34
   _____

4. 64
   − 58
   _____

5. 93
   − 27
   _____
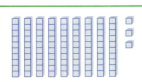

6. 68
   − 49
   _____
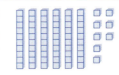

Subtract. Use coins if you like. To regroup, cross out 1 dime and draw 10 pennies.

7. 46¢
－ 25¢

8. 50¢
－ 17¢

9. 75¢
－ 38¢

10. 83¢
－ 55¢

11. 60¢
－ 43¢

12. 94¢
－ 58¢

★ PROBLEM SOLVING

13. Kevin would like to buy two model cars. Together they cost $20. One costs $2 more than the other. What is the price of each car?

_____

14. Kevin saw the cars on facing pages in a catalog, but he doesn't remember the page numbers. Can you help him? Here are some clues.

• The first page number is even.
• The second page number is odd.
• The sum of the page numbers is 69.

What are the two page numbers?

_____

That's AMAZING!

It takes 43 muscles in your face to frown but only 17 to smile.

How many more muscles does it take to frown than smile? (A face could get pretty tired frowning!)

_____

DO THE TWO-STEP!

Jackson stayed online for 56 minutes on Sunday. "That's too much computer time!" said his mom. So Jackson spent only 27 minutes online on Monday. How much less time did he spend online on Monday than on Sunday?

$$56 - 27 = ?$$

You know how to subtract with base-ten models. You can also subtract without them. Just follow these steps.

Step 1

Look at the ones.
Regroup if you need to.
Subtract the ones.

$$
\begin{array}{r}
{\scriptstyle 4\ 16} \\
\not5\not6 \\
-\ 27 \\
\hline
9
\end{array}
$$

Regroup 5 tens as 4 tens and 10 ones. Now you have 16 ones.
16 ones − 7 ones = 9 ones

Step 2

Subtract the tens.

$$
\begin{array}{r}
{\scriptstyle 4\ 16} \\
\not5\not6 \\
-\ 27 \\
\hline
29
\end{array}
$$

4 tens − 2 tens = 2 tens

Jackson spent 29 minutes less time online.

Your Turn

Estimate. Then find each difference. The first one is done for you.

1. $\begin{array}{r} 63 \\ -\ 41 \\ \hline 22 \end{array}$ $\begin{array}{r} 60 \\ -\ 40 \\ \hline 20 \end{array}$

2. $\begin{array}{r} 78¢ \\ -\ 9¢ \\ \hline \end{array}$

3. $\begin{array}{r} 42 \\ -\ 25 \\ \hline \end{array}$

4. $\begin{array}{r} 71 \\ -\ 17 \\ \hline \end{array}$

5. $\begin{array}{r} 50 \\ -\ 43 \\ \hline \end{array}$

6. $\begin{array}{r} 85 \\ -\ 19 \\ \hline \end{array}$

7. $\begin{array}{r} 73¢ \\ -\ 42¢ \\ \hline \end{array}$

8. $\begin{array}{r} 96 \\ -\ 28 \\ \hline \end{array}$

9. $\begin{array}{r} 47 \\ -\ 24 \\ \hline \end{array}$

10. $\begin{array}{r} 61 \\ -\ 38 \\ \hline \end{array}$

11. $\begin{array}{r} 95 \\ -\ 24 \\ \hline \end{array}$

12. $\begin{array}{r} 74 \\ -\ 56 \\ \hline \end{array}$

**Rewrite the exercise in vertical form.
Then subtract.**

TIP Estimate before or after you subtract. Then use your estimate to see if your answer is reasonable.

13. 76 − 34

14. $55 − $29

15. 70 − 43

16. $48 − $19

17. 66 − 22

18. 95¢ − 38¢

19. 85 − 48

20. 72¢ − 57¢

21. 43 − 25

22. Math Journal How do you know when to regroup to subtract? Explain. Give some examples.

★ PROBLEM SOLVING

23. Mrs. Bates wants to buy a bottle of ketchup. She has a coupon to help pay for it. How much will she pay if she uses the coupon?

_____

24. How much change should Ramón get if he uses 3 quarters to pay for the cabbage?

_____

89¢   57¢

COUPON 25¢ OFF

💡 Critical Thinking

Arrange these four digits in each exercise to get each sum or difference.

2 3 5 8

1.

+
8 1

2.
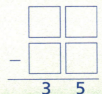
−
3 5

3.
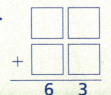
+
6 3

4.
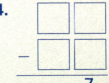
−
7

DO THE THREE-STEP!

Subtracting 3-digit numbers is like subtracting 2-digit numbers with an extra step. You have to subtract hundreds too.

Follow these steps to find $354 - 147$.

Step 1

Subtract the ones.
Regroup if you need to.

Regroup 5 tens as 4 tens and 10 ones.
Now you have 14 ones.
14 ones − 7 ones = 7 ones

$$
\begin{array}{r}
4\,14 \\
3\cancel{5}\cancel{4} \\
-\ 147 \\
\hline
7
\end{array}
$$

Step 2

Subtract the tens.
Regroup if you need to.

4 tens − 4 tens = 0 tens

$$
\begin{array}{r}
4\,14 \\
3\cancel{5}\cancel{4} \\
-\ 147 \\
\hline
07
\end{array}
$$

Step 3

Subtract the hundreds.

3 hundreds − 1 hundred =
2 hundreds

$$
\begin{array}{r}
4\,14 \\
3\cancel{5}\cancel{4} \\
-\ 147 \\
\hline
207
\end{array}
$$

You can add to check your answer.

$$
\begin{array}{r}
1 \\
207 \\
+\ 147 \\
\hline
354
\end{array}
$$

$354 - 147 = 207$

✐ Your Turn

Subtract. Add to check. Use base-ten models if you like.

1. $\ \ 265$ $-\ 127$	**2.** $\ \ 529$ $-\ 272$	**3.** $\ \ 437$ $-\ 309$	**4.** $\ \ 689$ $-\ 555$
5. $\ \ 395$ $-\ 280$	**6.** $\ \ 803$ $-\ 272$	**7.** $\ \ 692$ $-\ 437$	**8.** $\ \ 763$ $-\ 328$

Rewrite the exercise in vertical form. Then subtract. Add to check.

REMEMBER:
When you rewrite an exercise, be sure to line up ones below ones, tens below tens, and hundreds below hundreds.

9. 747 − 477

10. 963 − 457

11. 552 − 129

12. 309 − 63

13. 735 − 170

14. $429 − $53

15. $685 − $9

16. 340 − 27

17. 464 − 255

18. 786 − 696

19. $3.19 − $1.27

★ PROBLEM SOLVING

Write the correct number in each shape. (Hint: The same number is always used in the same shape.)

20. Which two numbers have a sum of 277 and difference of 1?

$$\boxed{} + \bigcirc = 277$$

$$\boxed{} - \bigcirc = 1$$

21. Which two numbers have a sum of 950 and a difference of 100?

$$\bigcirc + ✺ = 950$$

$$\bigcirc - ✺ = 100$$

💡 Critical Thinking

Look at how Jenna subtracted. What did she do wrong? Explain.

$$\begin{array}{r} 651 \\ -\ 426 \\ \hline 235 \end{array}$$

_____

_____

REGROUP THE GROUPS

Jasper was in school 179 days last year. How many days of the year was Jasper *not* in school?

To find out, subtract 179 from 365. That's the number of days in a year.

Step 1

Subtract the ones.
Regroup if you need to.

$$\begin{array}{r} \overset{5\,15}{3\cancel{6}\cancel{5}} \\ -\ 179 \\ \hline 6 \end{array}$$

Step 2

Subtract the tens.
Regroup if you need to.

$$\begin{array}{r} \overset{15}{2\,\cancel{5}\,15} \\ \cancel{3}\cancel{6}\cancel{5} \\ -\ 179 \\ \hline 86 \end{array}$$

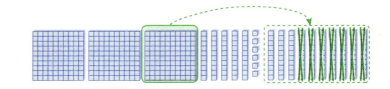

Step 3

Subtract the hundreds.

$$\begin{array}{r} \overset{15}{2\,\cancel{5}\,15} \\ \cancel{3}\cancel{6}\cancel{5} \\ -\ 179 \\ \hline 186 \end{array}$$

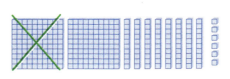

Add to check your answer.

$$\begin{array}{r} \overset{1\ 1}{186} \\ +\ 179 \\ \hline 365 \end{array}$$

TIP
You can also use a calculator to subtract. Estimate first so you'll have an idea of the answer in case you press a wrong key.

Jasper was *not* in school 186 days last year.

✏️ Your Turn

Subtract. Add to check. Use base-ten models if you like.

1. $\begin{array}{r} 942 \\ -\ 387 \\ \hline \end{array}$

2. $\begin{array}{r} 620 \\ -\ 105 \\ \hline \end{array}$

3. $\begin{array}{r} 356 \\ -\ 123 \\ \hline \end{array}$

4. $\begin{array}{r} 555 \\ -\ 297 \\ \hline \end{array}$

5. $\begin{array}{r} 692 \\ -\ 98 \\ \hline \end{array}$

6. $\begin{array}{r} 736 \\ -\ 375 \\ \hline \end{array}$

7. $\begin{array}{r} 423 \\ -\ 344 \\ \hline \end{array}$

8. $\begin{array}{r} 851 \\ -\ 667 \\ \hline \end{array}$

Cross out and draw base-ten models to find each difference.

9. Subtract 435 − 267.

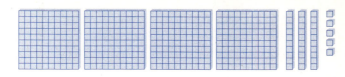

10. Subtract 713 − 599.

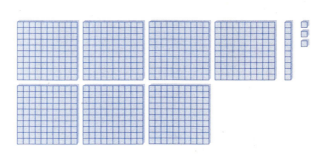

Look at all the exercises. Circle the ones that do not need regrouping. Do these first. Then do the others.

11.   571
   − 234

12.   825
   − 704

13.   154
   − 81

14.   569
   − 107

15.   482
   − 385

16.   782
   − 88

17.   566
   − 165

★ PROBLEM SOLVING

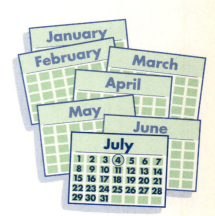

18. July 4th is the 185th day of a year. How many days are left in the year? _____

19. Jasper's mom works every day except on weekends and on 10 holidays. How many days a year does she work?
(Hint: There are 52 weeks in a year.) _____

 Critical Thinking

Estimate the difference. Then circle the correct answer.

1. 785 − 147   538 or 638

2. 745 − 187   558 or 658

3. 758 − 417   241 or 341

IT'S A FACT THAT IT'S EXACT!

Dara had a collection of 66 beanbag animals. She promised to give 38 of them to a children's hospital. "After all, I'll have plenty left," she thought.

How can you use mental math to find out how many she will have left?

Think 66 − 38 = ?

Here are some strategies you can use.

You can round to the nearest ten to get numbers that are easier to subtract.

Round 38 up 2 to 40. Then subtract.
66 − 40 = 26

Add back the 2.
26 + 2 = 28

So 66 − 38 = 28.

You can make a simpler problem by adding the same amount to both numbers.

Add 2 to both numbers.
66 + 2 = 68
38 + 2 = 40

68 − 40 = 28

So 66 − 38 = 28.

Dara will have 28 beanbag animals left.

Your Turn

Use mental math to subtract.

1. 57 − 30 = _____

2. 49 − 30 = _____

3. 156 − 80 = _____

4. 495 − 300 = _____

5. 653 − 550 = _____

6. 128 − 90 = _____

7. 899 − 490 = _____

8. 525 − 150 = _____

9. 1,555 − 750 = _____

Write <, =, or > in each circle.

10. 42 − 20 ◯ 20

11. 85 − 35 ◯ 50

12. 170 − 68 ◯ 100

13. 318 − 228 ◯ 100

14. 737 − 627 ◯ 100

15. 216 − 176 ◯ 50

Subtract mentally.

16. 76 − 43 = _____

17. 65 − 39 = _____

18. 476 − 43 = _____

19. 555 − 109 = _____

20. 607 − 305 = _____

21. 79 − 40 = _____

22. 65 − 43 = _____

23. 651 − 150 = _____

24. 508 − 79 = _____

25. 897 − 107 = _____

26. 175 − 139 = _____

27. 358 − 140 = _____

Find a number that you can add to both numbers to help you subtract. Write the number. Then subtract. The first one is done for you.

28. 43 − 19   Add __1__

    _44_ − _20_ = _24_

29. 62 − 47   Add _____

    ____ − ____ = ____

30. 485 − 158   Add _____

    ____ − ____ = ____

31. 133 − 89   Add _____

    ____ − ____ = ____

⭐ PROBLEM SOLVING

Use mental math.

32. Find the difference between 9,000 and 100. _____

33. Find the difference between 1,010 and 110. _____

One for fun!

WASHOUT!

Too many washings took the numbers off these shirts.
The sum of the numbers is 187.
The difference between them is 11.

Find the numbers. Write them
on the shirts so the game can begin!

FOUR STEPS FOR FOUR PLACES

Computer Town sold 1,908 computers the first year they opened. The second year, they sold 4,647. How many more computers did they sell the second year?

REMEMBER:
You can estimate or add to check your answer.

$4,647 - 1,908 = ?$

Subtracting 4-digit numbers is like subtracting 3-digit numbers but with an extra step. You need to subtract thousands too.

Step 1
Subtract the ones. Regroup if you need to.

$$\begin{array}{r} {}^{3\,17}\\ 4,6\!\!\!/4\!\!\!/7 \\ - 1,908 \\ \hline 9 \end{array}$$

Step 2
Subtract the tens. Regroup if you need to.

$$\begin{array}{r} {}^{3\,17}\\ 4,6\!\!\!/4\!\!\!/7 \\ - 1,908 \\ \hline 39 \end{array}$$

Step 3
Subtract the hundreds. Regroup if you need to.

$$\begin{array}{r} {}^{3\ 16\ 3\,17}\\ 4,6\!\!\!/4\!\!\!/7 \\ - 1,908 \\ \hline 739 \end{array}$$

Step 4
Subtract the thousands.

$$\begin{array}{r} {}^{3\ 16\ 3\,17}\\ 4,6\!\!\!/4\!\!\!/7 \\ - 1,908 \\ \hline 2,739 \end{array}$$

Add to check your answer.

$$\begin{array}{r} {}^{1\quad 1}\\ 2,739 \\ + 1,908 \\ \hline 4,647 \end{array}$$

Computer Town sold 2,739 more computers the second year.

Your Turn

Subtract. Estimate or add to check.

1. $\begin{array}{r} 3,479 \\ - 1,095 \\ \hline \end{array}$
2. $\begin{array}{r} 4,635 \\ - 3,708 \\ \hline \end{array}$
3. $\begin{array}{r} 5,678 \\ - 1,979 \\ \hline \end{array}$
4. $\begin{array}{r} 7,706 \\ - 2,462 \\ \hline \end{array}$

5. $\begin{array}{r} 6,750 \\ - 2,569 \\ \hline \end{array}$
6. $\begin{array}{r} 2,516 \\ - 1,815 \\ \hline \end{array}$
7. $\begin{array}{r} 1,234 \\ - \ \ 567 \\ \hline \end{array}$
8. $\begin{array}{r} 8,213 \\ - 3,405 \\ \hline \end{array}$

9. $\begin{array}{r} 9,452 \\ - 3,633 \\ \hline \end{array}$
10. $\begin{array}{r} 6,819 \\ - 2,547 \\ \hline \end{array}$
11. $\begin{array}{r} 7,056 \\ - 3,409 \\ \hline \end{array}$
12. $\begin{array}{r} 9,831 \\ - 6,525 \\ \hline \end{array}$

Look at all the exercises. Which ones look easiest? Do those first. Then do the others.

13.
$$\begin{array}{r} 5,429 \\ -\ 2,369 \\ \hline \end{array}$$

14.
$$\begin{array}{r} 4,700 \\ -\ \ \ \ 500 \\ \hline \end{array}$$

15.
$$\begin{array}{r} 8,215 \\ -\ 4,228 \\ \hline \end{array}$$

16.
$$\begin{array}{r} 7,118 \\ -\ 5,007 \\ \hline \end{array}$$

17.
$$\begin{array}{r} 8,674 \\ -\ 8,571 \\ \hline \end{array}$$

18.
$$\begin{array}{r} 4,532 \\ -\ 1,789 \\ \hline \end{array}$$

19.
$$\begin{array}{r} 7,304 \\ -\ 5,103 \\ \hline \end{array}$$

20.
$$\begin{array}{r} 5,240 \\ -\ \ \ 139 \\ \hline \end{array}$$

21.
$$\begin{array}{r} 3,297 \\ -\ \ \ 388 \\ \hline \end{array}$$

22.
$$\begin{array}{r} 7,450 \\ -\ 4,814 \\ \hline \end{array}$$

23.
$$\begin{array}{r} 8,133 \\ -\ 6,915 \\ \hline \end{array}$$

24.
$$\begin{array}{r} 9,412 \\ -\ 4,799 \\ \hline \end{array}$$

25. **Math Journal** What do you think is the hardest part of regrouping when you subtract 4-digit numbers? Give an example.

PROBLEM SOLVING

26. The telescope was invented in 1608. The magnifying glass was invented 358 years before that. In what year was the magnifying glass invented?

_____

27. Thomas A. Edison was named to the National Inventor's Hall of Fame in 1973. He'd invented the light bulb 94 years before. In what year did he invent the light bulb?

_____

That's AMAZING!

Inventors make, build, or design new things. How many years ago was each of these things invented?

1. the zipper in 1891
_____ years ago

2. pants suspenders in 1871
_____ years ago

3. the design of the Statue of Liberty in 1879
_____ years ago

THERE'S NOTHING TO IT!

The third graders are recycling plastic bottles. They sort them by the number codes on the bottoms of the bottles. The children sorted 403 bottles with . They sorted 129 with ②. How many more bottles with ① than ② were there?

$$403 - 129 = ?$$

Estimate first. $400 - 100 = 300$ The difference should be *close to* 300.

Step 1

Look at the ones. Regroup if you need to. If there are no tens, regroup a hundred.

```
  310
  4̸0̸3
- 129
```

Step 2

Regroup a ten so you can subtract the ones.

```
       9
  3̸1̸0̸13
  4̸0̸3̸
- 129
      4
```

Step 3

Subtract the tens. Finish subtracting.

```
       9
  3̸1̸0̸13
  4̸0̸3̸
- 129
   274
```

There were 274 more bottles with ① than ②.

That's close to the estimate of 300.

 Your Turn

> **REMEMBER:**
> You can also add to check your answer.

Estimate. Subtract. Compare your answer with your estimate.

1.
```
  806
- 268
```

2.
```
  350
- 178
```

3.
```
  600
- 453
```

4.
```
  506
- 347
```

5.
```
  708
- 239
```

6.
```
  200
-  67
```

7.
```
  902
- 785
```

8.
```
  408
- 279
```

9.
```
  401
-  92
```

10.
```
  802
- 619
```

11.
```
  500
- 248
```

12.
```
  908
- 539
```

Rewrite the exercise in vertical form. Then subtract. Estimate or add to check.

13. 308 − 126

14. 500 − 398

15. 5,050 − 4,389

16. 4,005 − 1,008

17. 2,057 − 88

18. 6,004 − 3,027

19. 1,205 − 9

20. 4,000 − 108

PROBLEM SOLVING

21. 708 people visited the Art Fair on Saturday. 538 came on Sunday. How many more came on Saturday than on Sunday?

_____

22. 239 of Saturday's visitors to the Art Fair were adults. How many were children?

_____

Critical Thinking

Ways to Subtract
- mental math
- pencil and paper
- calculator

Choose the best way to subtract. Tell why. Then subtract.

1. 500 − 6 = _____   _____

2. 4,238 − 2,762 = _____   _____

3. 700 − 102 = _____   _____

SPENDING MEANS SUBTRACTING

Chris was so hungry. She bought a big ham sandwich for $3.79. She paid for it with a $5 bill. How much change should she get?

$5.00 − $3.79 = ?

Subtracting money is like subtracting whole numbers. Just remember to write the dollar sign ($) and the decimal point (.) in the difference.

Estimate first. $5.00 − $4.00 = $1.00
The difference should be *about* $1.00.

You can also use a calculator to subtract. Be sure to press the ⬚ as you enter each amount.

Step 1	Step 2	Step 3
Write each amount. Line up the decimal points.	Subtract the way you would subtract whole numbers.	Write a dollar sign and a decimal point in the difference.

Step 1:
$5.00
− 3.79

Step 2:
   9
4 10̸10
$5̸.0̸0
− 3.79
1 21

Step 3:
   9
4 10̸10
$5̸.0̸0
− 3.79
$ 1.21

BEWARE!
You won't see a dollar sign on a calculator display. Just remember to write it in the difference.

Chris should get $1.21 in change. That's near the $1.00 estimate.

 Your Turn

Estimate the difference mentally. Then subtract. Don't forget to write a decimal point and a dollar sign in your answer.

1. $4.59
    − 1.75

2. $6.04
    − 0.98

3. $8.00
    − 2.39

4. $5.25
    − 3.66

5. $3.39
    − 1.83

6. $7.37
    − 4.09

7. $3.17
    − 2.98

8. $4.89
    − 1.98

9. $6.05
    − 3.09

10. $9.16
     − 5.57

Estimate mentally. Then subtract.

11. $8.02
    − 5.82

12. $7.95
    − 4.30

13. $5.00
    − 4.38

14. $4.50
    − 1.59

TIP
You can use play money to help you subtract real money. Pretend you are making change.

15. $20.00
    − 6.79

16. $15.00
    − 14.36

17. $62.00
    − 56.98

18. $50.00
    − 8.88

Problem Solving

Hungry? Use the menu to order some answers!
Use a calculator if you like.

19. Jake has a $5 bill. Is that enough for a turkey sandwich and 2 glasses of milk? Explain.

_____

_____

THE LUNCH MUNCH
Menu

Sandwiches
    Ham $3.75
    Turkey $3.50
    Cheese $1.69
    PB & J $1.35

Drinks
    Milk $0.65
    Juice $0.89

20. Andy wants to buy a cheese sandwich and a bottle of juice. How much change should he get if he pays with a $10 bill? How much change from a $100 bill?

_____

Critical Thinking

3 4 6 8
  5 7

Choose from these digits to solve each problem.
Use any digit more than once or not at all.

1. Write two 3-digit amounts with a difference of about $2.00.

2. Write two 4-digit amounts with a difference of about $15.00.

1. SAME ANSWERS

Write the correct number in each shape. (Hint: The same number is always used in the same shape.)

150 + [   ] = ⬭

250 − [   ] = ⬭

2. Pattern Differences

What if the pattern continues? What will be the difference between the first and last numbers in Row 10?

Row 1	50
Row 2	100 150
Row 3	150 200 250
Row 4	200 250 300 350

3. Picture This !

Kate's photo albums have a total of 97 pictures.

Three of these albums are Kate's. Circle her albums.

4. Find Me With Numbers

Jason gave Matt these directions to his house on Landon St. But he didn't give him the house number. Follow the directions to find Jason's house number.

- Start with 95.
- Add 3 thousands.
- Add 2 hundreds and 8 ones.
- Subtract 5 hundreds.

Jason lives at _____ Landon St.

5. Cross-Number Puzzle

Across

A. 39 + 43
B. 70 − 28
C. 152 − 78
D. 300 − 202

Down

A. 406 − 319
B. 700 − 658
C. 271 − 199
D. 68 + 26
E. 5,000 − 4,942

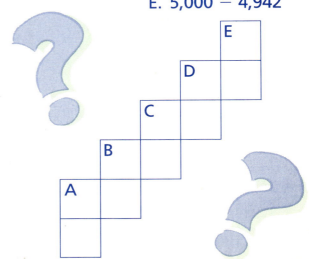

PLAY A  GAME

Get closest to 20!

Number of Players: 2 to 4

Goal: To score points by getting a difference close to 20

Materials:
- one spinner divided into 8 sections marked 0, 1, 2, 3, 4, 5, 6, 7
- paper and pencil for each player

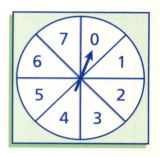

Game Rules

1. Players each spin the spinner 4 times and record the numbers they spin.

2. Each player uses the numbers to make two 2-digit numbers with a difference of 20 or close to 20.

3. Players compare their differences.

4. The player with the difference closest to 20 gets 1 point. If there's a tie, each of those players gets 1 point.

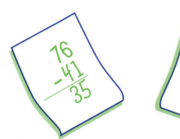

This player gets 1 point.

5. Players continue to take turns spinning and trying for a difference of 20.

6. The first player to get 10 points wins the game.

SHOW WHAT YOU KNOW . . . ABOUT ADDITION AND SUBTRACTION

Write each sum.

1. 7 + 7 = _____

2. 5 + 0 = _____

3. $8 + $9 = _____

4. 7¢ + 4¢ + 3¢ = _____

5. 6 + 5 + 0 + 5 = _____

Estimate the sum.

6. 37 + 52 _____

7. $69 + $25 _____

8. 219 + 411 _____

9. 479 + 166 _____

10. 2,331 + 956 _____

11. 835 + 5,123 _____

Use mental math to add.

12. 70 + 40 = _____

13. 300 + 500 = _____

14. 46¢ + 30¢ = _____

15. 45 + 19 = _____

16. $630 + $215 = _____

17. 59 + 43 = _____

Add.

18. 48¢
    + 34¢

19.   87
    + 65

20. $3.46
    + 1.54

21. 3,406
    + 955

Shade the oval for the correct answer.

22. 549   Ⓐ 512
    + 73   Ⓑ 612
           Ⓒ 622
           Ⓓ 722

23. 6,321   Ⓐ 7,000
    + 789   Ⓑ 7,010
             Ⓒ 7,100
             Ⓓ 7,110

24. $7.39   Ⓐ $8.26
    + 0.97   Ⓑ $8.36
             Ⓒ $9.26
             Ⓓ $9.36

Estimate to solve. Then explain your estimate.

25. Mr. Santos drove 421 miles from Lubbock to Austin. Then he drove 385 miles from Austin to Brownsville. About how many miles did he drive?

_____

_____

_____

Write each difference.

26. 8 − 0 = _____

27. 16¢ − 8¢ = _____

28. 12 − 3 = _____

29. 15 − 9 = _____

30. 9 − 9 = _____

31. $16 − $7 = _____

Estimate the difference.

32. 72¢ − 49¢ _____

33. 86 − 28 _____

34. $217 − $86 _____

35. 320 − 155 _____

36. 4,281 − 1,902 _____

37. 888 − 693 _____

Use mental math to subtract.

38. 90 − 50 = _____

39. $700 − $200 = _____

40. 1,400 − 600 = _____

41. 76¢ − 40¢ = _____

42. 928 − 300 = _____

43. $4.75 − $2.50 = _____

Subtract. Check your answer by adding.

44. 875
− 407

45. $6.00
− 4.38

46. 3,405
− 2,433

Shade the oval for the correct answer.

47. 607
− 438
Ⓐ 169
Ⓑ 179
Ⓒ 231
Ⓓ 269

48. $8.00
− 7.59
Ⓐ $0.41
Ⓑ $0.59
Ⓒ $1.41
Ⓓ $1.59

49. 7,003
− 475
Ⓐ 7,528
Ⓑ 6,528
Ⓒ 6,572
Ⓓ 7,472

Solve. Then explain how you got your answer.

50. Kyle bought some baseball cards for $3.98 and a cap for $5.79. How much change should he get from a $20 bill?

_____

_____

_____

THINKING ABOUT . . .
ADDITION AND SUBTRACTION

1 What did you learn about addition and subtraction?

_____

_____

_____

_____

2 What did you find easy about addition and subtraction?

_____

_____

_____

_____

3 What did you find difficult about addition and subtraction?

_____

_____

_____

_____

4 How do you use addition and subtraction in real life? Give some examples.

_____

_____

_____

_____

GLOSSARY

add To put two or more groups together. (p. 2)

addend The number that is added to another number. (p. 2)

$$3 + 5 = 8$$

addend

addition An operation for finding the total number when two or more groups are put together. Addition is the opposite of subtraction. (p. 2)

count back A strategy for subtracting 1, 2, or 3 from a number. (p. 12)

$$12 - 3 = 9$$

count on A strategy for adding 1, 2, or 3 to a number. (p. 2)

$$8 + 3 = 11$$

count up A strategy for subtracting numbers that are close. (p. 12)

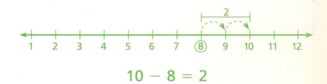

$$10 - 8 = 2$$

decimal point A symbol (.) used to separate dollars and cents in money amounts. (p. 40)

$2.70

decimal point

difference The answer in subtraction. (p. 12)

$$8 - 5 = 3$$

difference

digit One of the symbols 0, 1, 2, 3, 4, 5, 6, 7, 8, or 9 used for writing numbers. (p. 28)

dollar A bill or coin with a value of 100 cents. (p. 40)

dollar sign A symbol ($) used to mean dollars. (p. 40)

Write "3 dollars" as $3 or $3.00.

double Two addends that are the same. (p. 2)

      6 + 6 is a double.

estimate To find a number that is close to an exact answer. (p. 26)

fact family A group of related facts. (p. 18)

    3 + 5 = 8   8 − 5 = 3
    5 + 3 = 8   8 − 3 = 5

front-end estimation A way of estimating by using the digits in the greatest place-value position. (p. 42)

greater than (>) A relationship that shows that one number has a greater value than another. (p. 5)

  8 > 5 means *"8 is greater than 5."*

grouping property for addition The way addends are grouped does not change the sum. (p. 8)

    (3 + 8) + 2 = 13
    3 + (8 + 2) = 13

less than (<) A relationship that shows that one number has a lesser value than another. (p. 5)

  5 < 8 means *"5 is less than 8."*

making 10 A strategy for adding where addends are regrouped so that one addend becomes a 10. (p. 4)

    9 + 3 = 10 + 2
    9 + 3 = 12

mental math Finding an exact answer "in your head", using basic facts and strategies instead of using models, pencil and paper, or a calculator. (p. 35)

missing addend An unknown addend in an addition sentence. (p. 18)

$$8 + \bigcirc = 15$$

    ↑
  missing addend

number line A line with equally spaced points named by numbers. (p. 2)

number sentence A way of writing digits and symbols to show how numbers are related. (p. 19)

  2 + 9 = 11 is a number sentence.

order property of addition Changing the order of addends does not change the sum. (p. 6)

    8 + 2 = 10
    2 + 8 = 10

pattern A sequence of things that is repeated or which grows or shrinks in a certain way. (p. 22)

regroup To rename a number by exchanging 10 ones for 1 ten, 1 ten for 10 ones, 10 tens for 1 hundred, and so on. (p. 28)

Regroup 15 ones as 1 ten 5 ones.

related facts Facts that use the same numbers. (p. 16)

$7 + 4 = 11$ $11 - 4 = 7$

round To express a number to the nearest ten, hundred, or thousand. (p. 26)

823 rounded to the nearest ten is 820.

strategy A way of, or plan for, doing something. (p. 2)

subtract To take away or separate part of a group. (p. 12)

subtraction An operation that tells how many are left when one group is taken away from another. Subtraction is the opposite of addition. (p. 46)

sum The answer in addition. (p. 2)

$3 + 5 = 8$
↑
sum

zero property of addition The sum of 0 and any other number is always the other number. (p. 6)

$0 + 7 = 7$